THE COPPER WIRE VERSION OF ENERGY

Eva Prafulla Wankhede is a student at DPS Navi Mumbai, with a deep-seated passion for energy efficiency and sustainable living.

For a long time, Eva has been fascinated by the invisible forces that power our world, which has led her to merge this scientific curiosity with her love for writing.

Through her blog and now this book, she explores how technical energy concepts translate into everyday life and personal choices.

For Eva, this project is more than a publication—it is a significant step in her lifelong journey to explore, understand, and advocate for a more conscious and energy-efficient future.

THE COPPER WIRE VERSION OF ENERGY

Where Energy Meets Everyday Life

Eva Wankhede

Sakāl Publications

THE COPPER WIRE VERSION OF ENERGY

Sakal Media Pvt. Ltd.
595, Budhwar Peth,
Pune – 411002, India

www.sakalpublications.com
sakalprakashan@esakal.com

First Edition: May 2026

ISBN No.: 978-93-47138-11-9

Cover Design: Madhumita Shinde

Typesetting: Sakal Publications

Edited by: Lisa Braggs

Printed in India by Sakal Media Pvt. Ltd.

To the messy experiences that
lit the bulb,
and the good ones that
kept it lit

Contents

Finding my Frequency

Throughout my childhood, I've wanted to become a teacher as I loved teaching things to everyone. But never had interest in a specific subject. For me, anything that got me decent marks was my favourite subject, which was a constant loop-de-loop—the subject never stayed the same for a single semester.

It was weird to see people saying they loved a specific subject. For example, math, a subject which sits on a knife's edge of opinion. Never understood what was so fascinating about breaking your head over someone buying thirty-two watermelons or scribbling pages and pages to find that 'x' buried under a mountain of calculations.

And this was the same for me with all the subjects. But I never really bothered to find out why people liked a subject, how it applies to my life, or how it would in the future. I just knew I had to get good grades and somehow pass onto the next academic year, with a big chunk of academic knowledge wiping away the minute after finals. And there was nothing strange about it, since it was the same every year.

But it was never just studies. I've always had fifty other things going on along with it. From my childhood to my teenage years, I've tried out every possible thing—singing, dancing, painting, tennis, football, basketball, drama, and whatnot. I loved doing these so much that I don't remember a single year without them, except for the year of my 10th board exams where I was focused on academics—a story for another time.

Anyways, I did all this till about 8th grade and then it was time for high school, and *ohhh myy*... it left me speechless. It was as if someone threw me off a secluded corner of the planet Pluto which never existed. In the first few months, everything went past me.

To cope with the new academics, for the first time in my life, I had a dozen classes—not for extracurriculars but for academics. But now, every day was a repetition—school to home, home to class, one class to another class, class to home, and boom! It's a new day already, just to do everything all over again the next day. This was completely out of character. New world, new me—and I didn't like it.

Before high school, it was the constant change that I loved. The way everyday was a new day with new activities was something that kept me going. Somehow, I made it through that year.

Then, the year of the board exams arrived and it was time to select the stream I would pursue. Being affiliated with CBSE, we had a few options to choose from—Science (Medical and Non-medical), Commerce, Humanities, and Liberal Arts.

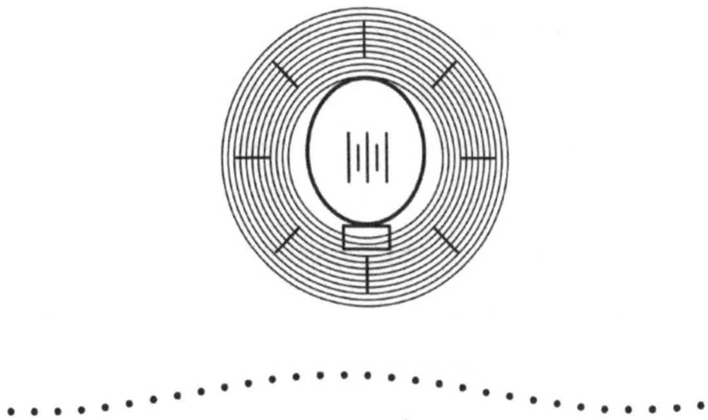

Finding the spark

It was clear to me that I wanted to choose science, not because I loved the subject, but rather I tried the process of elimination. Social studies, and those integrated in it like history, was hard for me. I was neither fond of arts or literature. I did like economics but to keep my options open, I went with science.

After my boards were finally over, I went back to explore activities and co curriculars. This was the period where I truly knew what I wanted to pursue and why. Basically, one thing was sure, whatever I choose to pursue would not involve me just sitting in one place all day long with the same repetitive schedule. To realise this, my 9th and 10th grade rang the bells.

Now, I was always surrounded by talks of energy, gas, heating, and stuff like that because of my parents being in that sector, which is why I briefly knew about it but it never really sparked an interest in me. It was during the holidays after my exams that I came across the research programme, NYAS. It required the students to submit applications consisting of essays and interests. The research was based on solving real world problems. It was April at that time and, unusually, it started raining frequently. The rains, which would usually come in July–August, greeted Mumbai in April, indicating that the global temperature was rising.

So, I chose this topic and presented solutions to curb it for my application process. Not knowing much about the topic, I went ahead and asked my dad if he knew ways to curb it and he introduced me to the concept of energy efficiency. It's basically maximising the output by minimising the input using suitable technologies.

This concept felt so relevant and easy to understand that it unexpectedly sparked curiosity in me. The reason? Probably because this was something I could relate to and apply in real life. And this is where I found my starting point. A few months into me learning about energy efficiency, it had me hooked.

While this was going on, I chose to volunteer for an event our school was participating in. The event was basically a counselling movement emphasizing the concept that every teacher is a counsellor. It was held for 3–4 days and required student volunteers, which was me and another student from my grade.

A few teachers had to go to a few schools to attend summits and more exciting stuff. The students were there to go around helping and weren't asked to attend the session meant for teachers.

What's so great about that? Well, though luck has rarely been on my side, this instance proved an exception. My friend and I were the sole student attendees at the session, surrounded by university mentors and teachers.

With only the two of us, the teachers had no choice but to keep us together with them. But guess what? The 'teachers only' session was actually a secret career counselling for us clueless students. Never in a million guesses did I think I'd walk out of there with a direction.

The speaker was Aditya Jhunjhunwala. And what he talked about in the session was something that I had never heard of before.

From the moment a child takes its first breath into this world, specifically India, the first line from families and society is "Hey, welcome! Here's a shiny title—*"engineer"*. I'm not saying it's the same for every family, but for a large portion, it goes like this. So, engineering is a profession we've all heard of throughout our life. But in this session, I was introduced to the concept of *imagineers*.

The speaker described how engineering is like a seven-course meal—It has to follow the exact same path of education, same way of thinking—which limits us from thinking beyond. While being an *imagineer* is like a traditional Indian *thaali*, there's no 'first-this-then-that' rule.

You can toss together random food items and yet it turns out amazing, and, personally, a lot more savoury. It basically gives you the freedom to mix and match anything with anything at all, unlike the seven-course meal.

This concept really stuck with me, cleared doubts I never had about myself, gave me a direction to think and, most importantly, brought me personal relief regarding my dislike for following a specific pattern or going through the loop-de-loop. It just lifted a weight I didn't even know I was carrying all along. I will always be grateful to my teachers who decided to take us to the session.

Now, I was pretty much sure that I wanted to enter into the energy sector, and the one way to get involved was engineering. For me, I always preferred doing work or any activity which required me to move around loads, interact with people, come up with my own ideas, and integrate them into anything around me.

So, now I'm considering going into something like mechanical engineering, mechatronics along with energy. Having always preferred hands-on learning and work in general, it's a win for me since the energy sector has just that.

My interest in this field also developed when I participated in my school's Budding *Scientists* competition and later, the research programme the same year. Also, I loved how there wasn't much competition involved, unlike competitive programmes like computer science or professions like being a doctor, cause I'm a lazy person. Unless I truly like something, it's hard to get myself started on anything.

By the way, medical was never for me.
First, I never understood biology, and had no interest in it at all.

Second—blood? needles? *Nah.* I always freaked out looking at the dissection videos our biology teacher showed us.

Third, as I stated, I'm not a bookworm who can sit for hours at the desk endlessly memorising things and having the patience to study a mountain of course material.

It was like forcing me to run a marathon in flip flops while I'm half asleep. I hold immense respect for those who genuinely love biology and devote hours to study it. They're basically a different, super powered species of humans.

That concluded my questionably long hunt for a career path.

Global Warming—Energy Efficiency

These days, seasons, weather, and climate are constantly shuffling. One day, you're in a desert, the next, you're dodging raindrops in monsoon. Be it anywhere in the world, every passing day, we all can see a significant rise in the earth's temperature.

In my locality, the monsoon usually begins in June, but for the first time, it started in May, which was alarming. In the UK too, a noticeable change has occurred, with hotter summers and temperatures exceeding 40 degrees Celsius.

And who's stirred up this chaos? An answer that's, like, no surprise—global warming. What exactly is it? Global warming is rapid climate change, which isn't a normal thing. Climate change is normal and happens every 30–40 years—nothing concerning. But it's the 'rapid' that makes it abnormal and threatening.

There are lots of negative effects of global warming—like the average temperature rising, melting ice glaciers adding to rising sea levels, effects on biodiversity—causing humanity to suffer with food, water, houses, agriculture, and various diseases.

The reason for this is the emission of harmful gases, mainly CO_2, into the air due to an increase in industrialization over the past years. This happens due to burning of fossil fuels, mainly coal, which release greenhouse gases. These are also non-renewable sources—they can't be replenished at the rate they are consumed.

A warming world needs cooling solutions

The planet won't let us take a million-year-chilling nap, hoping for another batch of dinosaurs to magically appear again and fossilize for us. We need to cut down the emissions today, not plan a fossil-hunting expedition in the far future.

One present example of a country facing severe problems due to global warming is a tiny, beautiful nation in the South Pacific Ocean—Tuvalu. It's extremely low-lying with its highest point being only 15 feet above the sea.

This country is on the verge of disappearing due to rising sea levels caused by global warming. As the Earth gets warmer, glaciers melt and ocean water expands, causing the sea to rise. Scientists predict that by 2025, half of Tuvalu's capital, Funafuti, could be flooded by tides.

To save their people and culture, Tuvalu is planning steps to make it a digital nation. They are using 3-D scanners to scan their islands, recording their traditional songs, and making apps to move government services online.

Even if the physical land is gone, their history and digital country will still exist. But while failing to save the hardware version (the land), can we really save the software version (the digital nation)?

Along with this, they also signed a historic treaty with Australia in 2023. It allows up to 280 citizens of Tuvalu to move to Australia every year—giving rise to the world's first climate crisis visas, making them climate refugees. They get the rights to live, work, education, healthcare facilities, and more.

Tuvalu, being so small, might be the first one to show the world what will happen if global warming isn't stopped—with islands like Maldives, Kiribati, and Marshall Islands forced to stand next in the line.

Even with the weather escalating to this magnitude, it's still possible to tame the situation. This isn't the first time the world has faced a critical crisis like this.

When an ozone hole was found in the 1980s above Antarctica, the entire world came together—found the problem, understood the situation, had discussions, meetings, and successful solution execution, all in about 10 years—quite hypothetical to imagine this in 2026, seeing the ongoing seamless wars.

But yes, the people of the 80s were apparently a lot more united and concerned with actual issues. This was probably one of the greatest achievements in human history.

Some traditional, textbook-like solutions to reduce global warming would be to increase the use of renewable energy sources, carbon and methane capture and sequestration, land use, and reforestation. But these aren't immediate solutions and will take years to develop and come into effective use.

The more suitable way to counter this crisis, I believe, would be 'Improving energy efficiency'. This uses less energy to attain the same outcome. It's about maximising the output of a system while minimising the energy input.

This reduces energy waste, mainly CO_2 emissions, and benefits the environment instantly. This can involve an increase in use of energy efficient appliances.

Hot water boilers and air-conditioners can be replaced by a single, most efficient equipment, the 'heat pump'. Or take the example of energy star rated appliances. They are designed to use less energy but give the same output.

The evolution of the light bulb is also associated with this. Moving from incandescent bulbs to fluorescent bulbs and ultimately LED technology. LED bulbs are 90% efficient, CFL bulbs are 85% efficient, while incandescent bulbs are only 10% efficient.

The incandescent bulbs are just heat generating devices that accidently make some light—kind of like a by-product released in a chemical reaction. LED technology has emerged as the most energy efficient option since it consumes less energy and lasts much longer compared to the previous ones.

It's a technology which can be implemented immediately, that too at a reasonable cost. Great for the current world, who, after the pandemic, has run out of patience for anything that lasts longer than a 30-second reel—scrolling our way through civilization.

We're a species that can't sit through a long-form video, let alone wait for a 50-year plan. We've lost our patience, so the technology better keep up.

While long-term solutions are vital, enhancing energy efficiency represents an urgent and practical step we can take now to help mitigate the adverse effects of climate change.

Change is the Constant

Can you imagine going from India to London on a horse cart in 2026? Or the recent thing, comparing 2020 and 2026. The things which were normal in 2020 have become the horse carts of today. In today's world, or at least in my locality, heavy rains are no longer an excuse to skip school. Your classroom is where your device is.

Ohhhh the good old days, where holidays meant actually getting a break and not having to stare at the laptop screen as compensation.

All this shows that change is inevitable. Just like the saying goes—"Life is like a comedy show where the punchlines keep changing, but the audience is always expected to laugh".

Change is the constant; it is something the great scientist, Darwin, spent most of his life working on. Darwin's core insight was that the environment is never static or still.

Climate changes, living species disappear and evolve, food sources modify and vanish. If a species stops changing while its environment continues to change, it faces extinction.

Darwin's principle gives the idea that what refuses to break will bend and what learns to yield will endure—meaning, "it's not the strongest of the species that survives, nor the most intelligent... it is the one that is most adaptable to change", quotes the business professor, Leon C. Megginson. He was paraphrasing Darwin's ideas to explain how businesses need to change to stay alive.

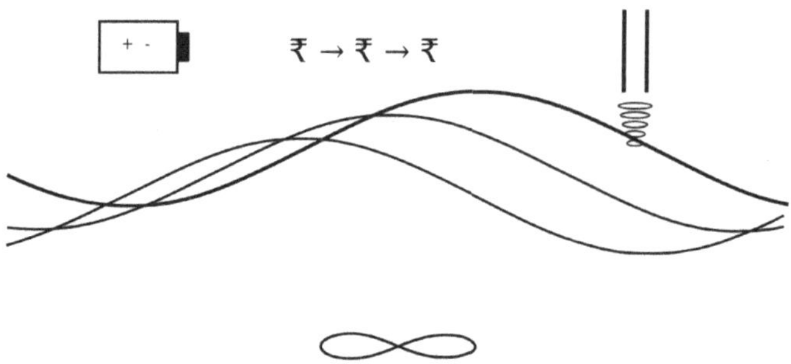

Everything flows, nothing stands still

Even in the world of energy, change is the only factor making stuff work. If energy stays still, it's of no use. This can be seen in batteries, waterfalls, dams—which work on the concept of waterfall, wind, and a lot more.

In the case of a battery, it only works due to the constant change of its internal chemicals into electricity. The battery wants to move its energy from the positive side to the negative side. If the movement which causes the change stops, the battery dies.

And, in the case of a waterfall, water at the top that is still, has the potential to create energy, but it only generates electricity during its fall—same in the case of dams.

Now, another notable thing in terms of energy is the law of thermodynamics, which states that energy can neither be created nor be destroyed and can only be transferred from one phase to another.

The first law of thermodynamics is proof that change is constant. If it couldn't change its form or phase, the universe would be a still picture where nothing ever happens. Some common examples of this are cars moving, plants growing, and photosynthesis.

This law doesn't just apply to water and electrons, it's also the logic behind the economics that runs the world, which is the cash flow. Cash flow is the movement of money—it is the amount of money coming in (cash inflow) and going out (cash outflow). And the amount left after subtracting the outflows from the inflows is what gives us the net cash flow.

If one just keeps their money still and doesn't make it flow constantly, it loses its purpose as a tool for growth and survival. Still money loses its purchasing power over time due to inflation.

Robert Kiyosaki, the author of *Rich Dad Poor Dad*, believes and argues that focusing on cash flow is what leads to wealth, not just a high salary. Along with this, research also consistently shows cash flow as one of the leading causes of failed businesses. About 82% of small businesses fail due to poor cash flow management.

Money isn't just a thing to be kept locked in a highly secure, 10-digit password protected safe box, rather it's the current which should be flowing constantly. Just like electricity flows to do work, money must also flow constantly to create value.

All this shows that 'Change is the Constant' is the underlying principle of all things. Be it biology, philosophy, physics, chemistry, economics—you name it and its there. It's all around us, and it's what our universe functions on.

To sum it all up, in the world of energy and machines, a system which has stopped working is of no use. The same applies to us. Adaptability isn't just a skill which is nice to have but rather a necessity.

We live in 2026, where our tomorrow, our future and

our map, all are constantly being rewritten by climate, technology, and new emerging ideas along the way. In a constantly changing and moving world, the only way to stay balanced and on track is to keep flowing.

Communication as Efficiency

The celestial expanse, resplendent with ethereal beauty, presents itself to our gaze, rivaling natural splendor. This sublime display evokes an affirmation of its captivating allure.

You know what this means? Simply, the sky is pretty. The above lines are exactly what I'm expected to write in a test. Because apparently, to sound clever, just complicate things unnecessarily.

Clever = Using 10 complex words when 2 will do. This is something I never understood. Going back into the history of evolution, communication was invented to convey our feelings or message to the other person. This was

probably the primary reason. So, is it really necessary to sound fancy?

Whenever one uses William Shakespeare typa vocab, by the time the listener finishes decoding the complicated sentence, the original spark of the idea loses its power. While simple words are like a high-quality copper wire—the message travels at the speed of light with zero loss.

Usually, we're taught that a good essay is the one that requires a dictionary to finish. If it's not confusing, is it even an essay? We're graded on how many syllables we can cram into a sentence. Using 'resplendent' instead of 'pretty' somehow makes the idea 10x more powerful. But, from my point of view, this is a massive energy leak.

In the world of machinery, my goal is to get from Point A to point B as efficiently and accurately as possible. If I'm building a circuit, I don't want a fancy looking wire that looks pretty but loses half the electricity to heat. I'll always look for the most efficient path for the current to flow.

Communication is no different. When we use hard vocab just for the sake of it, we add mental resistance and make listeners work ten times harder to find the actual message buried under a mountain of fancy adjectives.

I've realized, the smartest people aren't the ones using the biggest words—they are the ones who can compress a heavy concept in a few words, so it flows perfectly into anyone's mind without blowing a fuse.

Efficiency in communication is Output divided by Input. If you use 100 difficult words to tell me something that can be said in simply 5 words, your efficiency is only 5%. An *imagineer* or any technical professional doesn't want a 5% efficient engine. You want the most understandable output while minimizing the word input. Using short and simple words isn't being basic, it is being energy efficient.

This, I believe, also relates to teaching. My liking for energy efficiency is due to my dad explaining the concept in such easy words that it's hard to not get hooked on it. He didn't use a mountain of academic adjectives and technical stuff to explain it; just the copper wire version of the truth. Because of that, the bulb in my head didn't just flicker; it stayed lit.

Another incident is that of physics. Till 10th grade, physics was quite an understandable subject, but transitioning to 11th was like a 100-foot drop into a dark pit of Greek letters, where k is the constant for everything.

10th grade physics is simple and polite—just plug in two numbers in a 4-letter formula, get the right answer, and walk away feeling like the next Newton. But then, 11th grade happens, and it's as if the 'x' you were looking for in math migrated to physics, and it brought along its entire extended family—alpha, beta, gamma, and so on.

You start solving the question on page 1, and by the time you reach the final answer it's already page 4. By then, you've even forgotten what the original question was.

You go from finding "How fast is the car moving?" to "If a ball is sliding down a frictionless incline plane at a 30-degree angle and the wind is blowing at 10 m/s and the observer is rotating on a merry go round... find the tension in the string that shouldn't even exist".

So, for me, 11th grade physics was confusing and frustrating. Learning from textbooks and lectures was like trying to

swallow a 6-piece pizza in one go. I tried online fixes and tutors, but my brain went, “Nope, still confused”.

Then, a month or two before my finals, I found a great teacher, Anand Sir. While others were busy with complex stuff, he simplified it—“it’s just stuff around you”. And when he broke down the tech-speak my brain went—“Ain’t so bad”. Academic physics started making so much sense.

In the end, every conversation is a transfer of energy. Using basic, easy to catch words light a spark in our brain. It’s fast and it’s 100% efficient. And when you are wrapping your head in complex vocab—you aren’t creating a spark but rather smoke.

That smoke is the mental friction of the listener trying to make sense and catching the adjectives to find the point of the conversation.

In machinery, friction is an enemy—it wears and tears things, wastes power, energy and creates unwanted heat. Why would we want that in our minds?

Overflowing with Potential Energy—BEING BORED

I love being bored—sitting down, doing absolutely nothing, with no thoughts and an empty mind, disconnected from the world around me. It seemed kind of weird at first but now, it's clicked for me. Why and how did I develop a liking for it?

Well, I never really had any social media which, at first, felt like I was in a parallel world where only I existed, stuck in a forgotten corner of the universe.

All thanks to COVID, it was as if overnight, people went from asking names and phone numbers to asking Insta IDs to befriend or acknowledge someone's presence. Every time I interacted with someone or had a mere 'hi hello' chat with those in my grade, their first question would be, "What's your Insta ID?"

Well, this question always annoyed me. It was kind of awkward admitting I didn't have Instagram—it felt like I was missing out on something everyone else was on. This is what having FOMO—Fear Of Missing Out—felt like the first time around.

Why, you say, did I not have any socials after this? Well, my parents were always clear about one thing—no social media. And I, too, didn't try asking for it since I knew their reason for keeping me out of it.

I knew the consequences if I didn't use it properly. So, with a heavy heart, I decided to not have it. Like I said before, it was weirdly deflating. And I felt that for about two years, until our school decided to have a week-long picnic.

This was the first time our school had such a long trip, since the previous trips were usually limited to a single day. So obviously, I was very excited! But I definitely did not expect that things might take a turn for the worse.

I spent the trip happily and enjoyed every moment since I only tagged along with my best friend, who also didn't have social media. But, for the majority of others the whole trip was a disaster, you know, the kind of events people would tell stories about for years while remembering the questionably good old days.

To tell you the root cause of it, it was all social media. Without getting into the details—since I didn't pay much attention and forgot about the incident soon after(I have a terrible memory)basically, a bunch of people said something about another bunch of people on social media.

Soon, Instagram turned into a battlefield—everyone fighting, no one offline.

And I'm lowkey a lazy person, whenever things get too complicated and I know racking my brain over it won't be worth it, I distance myself from the chaos and don't process it at all. Seeing how tense everyone was for literally no good reason over mere internet comments, for the first time, I felt happy about not being hypnotised by the Insta culture.

This was just one of the very few incidents, and as time went by, after observing most people around me, I finally understood the genius behind my parents' and my decision to abstain from socials.

Another thing I noticed was how my peers were slowly losing their own individuality. Everyone had the exact same personality, behaviour, attitude, aesthetics, and more. Due to the so-called 'Influencers', everyone has been hooked by the glam world.

Originality is so last season because, apparently, the latest trend is copying your fav influencer's vibe and their questionable lifestyle choices. Meet ups are basically a copy-paste station—everyone trying to recreate the same viral picture. Birthday parties are like a fashion show of clones—same pose, same outfit, even the same smile.

Seeing how invested everyone is in some other, third person's life, you leave no space for the brain to come up with thoughts and ideas of your own. It's constantly crowded with thoughts of other people. It was quite easy to notice this drastic change everywhere since I always have had enough space in my brain to process and come up with thoughts.

This is what developed the 'JOMO'—Joy Of Missing Out—for me. I love my own company and spending time with the usual 3–4 people, untouched by the ongoing chaos.

This is when I realised how much I love being bored. Whenever I sit down with nothing, it helps me come up with the best ideas and thoughts. I usually do this when I plan to start writing a blog, make a science project, or simply strategize how I'll get through my dreadful midterms or annuals.

Being constantly reserved by something restricts our brains ability to come up with something different and guide us to think in a particular direction—which is heavily influenced by what we see around us, for most being social media reels. Only when we aren't mentally caged can we actually think of the most surreal of things. Like it's said, imagination has no limits, and it's true. But we can surely narrow it down to something equally brilliant but realistic.

This also gives rise to the term *imagineers*, a combination of engineers and imagination. For a while now, the world has been busy making engineers who are taught the same course and are given the same direction to think.

But now, the need of the hour is *imagineers,* who go out of their way and bring ideas to life which would otherwise seem unreal

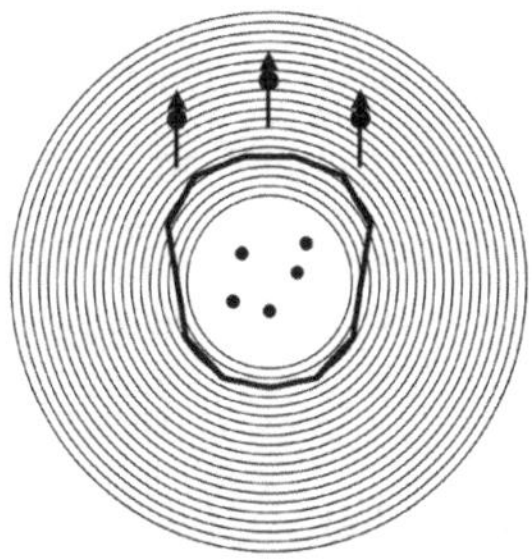

An empty mind overflows with potential

when just talked about. This, I think, will happen when we start getting bored and have zero thoughts to start with.

Whenever I mention this, people often mistake it for overthinking, but it's different. Overthinking is usually dwelling on the past events—which probably doesn't matter in the present—or worrying about the future. What I'm actually referring to is a state of not thinking about something specific, its more like a blank or neutral state.

It's not about overanalysing or reassessing things, it's just a lack of focused thoughts. This is also one of the reasons why people with bad memory are usually the happier ones since they easily forget the past and have nothing to overanalyse. Who needs to remember everything? Less memory, less stress.

Now, getting back to being bored. When we look at our past, a few decades or centuries ago, we had one of the best inventions and crazy, creative stuff come to life that are still being used to date. That's probably because nothing kept people reserved all the time, unlike today with social media, where the brain needs continuous dopamine hits to get through the day.
For instance, take the most well-known example, Newton under an apple tree. The apple fell, Newton freaked out, and physics was born. It was a game changer for physics but a recurring headache and nightmare for students. Love it or hate it, it's a pivotal moment in science history.

However, if Newton was born in today's generation and would've been sitting under an apple tree, there's every possibility that he would've simply ignored the fallen apple and continued scrolling. But, by him not having fifty tabs open in his brain all the time, could he have actually been able to think why the apple fell and figure it out how it happened.

Few more examples are that of James Watt's Steam Engine, where he was simply staring at a boiling kettle when he realised steam could power engines. This sparked the Industrial Revolution, transforming industries and transportation.

Also, Albert Einstein's Theory of Relativity, where Einstein was daydreaming about chasing a beam of light when he developed the theory of relativity, revolutionizing our understanding of space and time.

To sum it up, we should start making our brain overflow with potential energy. Being bored isn't the end of the world, it's a chance to reboot, find happiness and surreal ideas. Start embracing those blank moments.

The Art of Unplanned feat. Dora The Explorer—The Famous Cartoon

Dora the Explorer—Ohh, this lost-but-proud explorer who's got a PhD in asking the obvious—"What's that thing right in front of me?", was one of the cartoons I binge watched everyday for who knows what good reason.

Dora is the Itinerary Icon, the go-to guru for people who cannot walk two steps to the fridge without a five-step checklist. In the show, she literally cannot move without her map and a backpack that has everything for every possible scenario.

And it's not only Dora that follows a 20-step to-do list every day, I'm sure most of us would have tried doing this once. But in real life, being 'Dora' is exhausting. Constantly checking the map—which in our case is a schedule or the GPS—is high resistance activity. It burns the mental battery just to make sure you're on the right path.

This, obviously, does not apply to a factory where a glitch is a disaster. For us humans, a glitch is where the magic happens. Being strategically unplanned means you have a goal but aren't wearing blinders that keep you from seeing the real scenery.

Now, I'm not saying one should be living life as a directionless glitch. You need a goal. You need a destination. But the problem starts when we become so absorbed in the itinerary that we treat a day like a high-stakes Subway Surfers mission. We become so focused on the map that we forget that the real reason we stepped out is actually to explore.

You have your goal, but you leave enough slack in your mental battery to end up taking the long way. Real discovery and exploration don't happen when you're staring at the blue line of the GPS, but rather in the unplanned gaps between the points on your map.

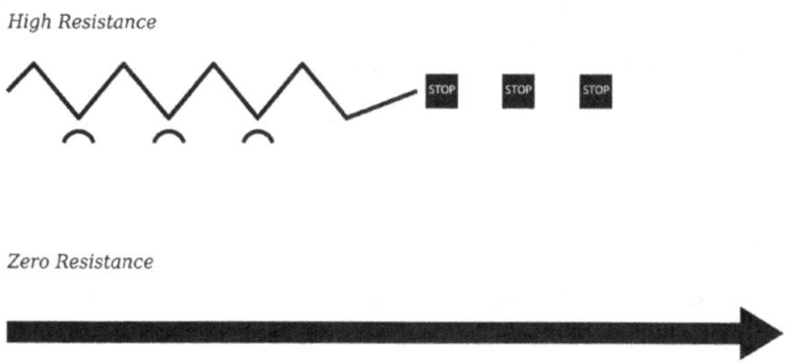

Be the superconductor, not the toaster

When you stop obsessing over the right way to do things is when you leave room for serendipity—a fancy word for happy accidents. You never know which bulb in your brain it'll light up.

Rather than carrying around every app, every notification, and every just-in-case plan, why not have an unplanned day by travelling light. You don't need a backpack of digital tools when you trust the flow of current. You save energy by not preparing for a thousand what-ifs.

In Dora the Explorer, Dora is a high resistance machine. In simple words, resistance is anything that slows down or blocks a flow. One of the things it is used in is toasters.

Resistance in a toaster is like a deliberate electrical traffic jam. As the electricity tries to pass through the wire, it constantly rubs and bumps against the sides. This electrical friction creates the heat that toasts the bread.

Now, Dora is high resistance because she can't just explore. She has to stop, consult a map, sing a song, and then wait for the manual input from the audience.

All that stopping and starting is just friction that turns her energy into useless heat, which we can say is wasted time that could have been used to cover the actual distance instead. This might be a fine approach for maybe 5 out of 8 days, but those unplanned 2 days are what click.

She's the perfect example of why a lot of us are always tired—we are burning all our energy trying to be on track instead of just being superconductors who let the day take us where it wants.

It's about knowing when to be a programmed machine and when to be a superconductor.

A superconductor is like a frictionless, wide-open highway for electricity to pass through. These allow energy to flow with zero resistance. This means no energy is wasted as heat and current keeps moving without slowing down.

In our context, being a superconductor means to stop bumping into schedules and getting lost using maps, but rather making the day flow efficiently. It's like putting your brain in power saving mode.

If I were to relate it to something, I'd take the example of Newton. While sitting under the tree, he wasn't planning on inventing physics that day, he was just having an unplanned moment under the tree, which, out of nowhere, rang the bells for him.

Maybe we all need to be a little less like Dora on a daily basis, and be more like a superconductor on some days—embrace the art of being unplanned little by little. We need to stop fighting the friction of our own schedules and let the day flow through us.

Real energy efficiency isn't about how well and quick you follow the map—it's about lowering your resistance, so you generate the potential energy to catch the apple the moment it falls.

Pressure Cooker

When life gives Asians lemons, they make lemonade. But first, they need their rice. Rice is the Asian version of "just in case"—just in case of hunger, just in case of an emergency. You'll find it in every meal of theirs.

But ever wondered which mastermind thought, "Lets trap some steam and cook food faster"? Well, Asian kitchens send a thank you note to the inventor of the pressure cooker—you've single-handedly reduced cooking time by decades.

So, the inventor who brought us this amazing gift is Denis Papin, a French physicist. He has not only cut down the cooking time but also helped in saving tonnes of energy. He wasn't just trying to cook faster; he was a scientist obsessed with the power of steam.

He realized that if you seal a pot tight, the steam can't escape. This creates pressure, which forces the water to stay liquid even at temperatures up to 121 degrees Celsius.

Why is a pressure cooker better than a normal pot? Well, in a normal pot, the energy is continuously wasted into the kitchen air in the form of vapour. When water boils in an open pot, it turns to steam. And to turn that liquid into gas, it must absorb energy 5.4 times more than what it takes to just raise the temperature to the boiling point.

While a pressure cooker is a closed loop system. To explain a closed loop system, it's a circular process where resources are reused and recycled with little to no waste. This system minimizes waste and external inputs needed.

So, in this case, by preventing evaporation, fuel isn't wasted on the phase change which doesn't play a role in the actual cooking process.

Now, inside the pot, the cooking happens using the potential energy stored. The steam molecules are constantly causing chaos—bouncing, going crazy, plotting their great escape from the cooker. This brings the whistle (the weight) into picture. It's an important safety valve that lets out just enough kinetic energy, so the vessel doesn't explode—if the pressure in the cooker is greater, lift the safety valve.

By using a pressure cooker, a household can save up to 70% of energy costs compared to traditional cooking method like boiling or using an oven. But this doesn't just cut down cooking time and reduce energy consumption, it also maximizes the output from the ingredients, making it fully energy efficient.

As pressure cooking uses very little water and finishes in a jiffy, it retains up to 90–95% of nutrients that usually escape into the water when boiling. Also, for families using limited or traditional biomass fuels, a pressure cooker doubles the life of their fuel source.

And now, there exist the modern electric pressure cookers which have built in microprocessors. These are independent automated devices that track the temperature and automatically turn the heating unit off once the desired pressure is reached, maintaining the heat using only 60% of the cooking time in actual electricity.

There's one more thing that really elevates the use of pressure cookers. Think of the pressure cooker as a sterile vault—a germ free safe for your food. During the high voltage cooking phase, the extreme heat acts like a reset button, killing the bacteria which is responsible for food spoilage.

When the cooker cools down, the steam condenses (turns into water), creating a vacuum that sucks the lid shut with a massive force. This airtight seal acts as a freshness lock—keeping new germs and oxygen out to prevent oxidation and other bacteria formation while perfectly preserving the food item for 2-3 days even without a fridge. It pauses the soiling process the moment the whistle stops.

Now, let's talk about the economics of pressure cookers in terms of OPEX and CAPEX.

OPEX—Operating expenditure—Entry Fee
CAPEX—Capital expenditure—Subscription fee

Talking about CAPEX, the pressure cooker has a very low CAPEX.

It's a one-time investment on a durable piece of metal. It doesn't require some high-tech machinery to build the cooker or some specialized technician to fix it. With a small investment and a one-time payment, you own a machine that can last up to 20 years, saving huge amounts of energy and time.

Now, to talk about the OPEX for a pressure cooker, it's the lowest in the kitchen.

Since it cuts down cooking time, your monthly fuel requirement, which can be LPG or electricity, drops significantly. You are paying for a machine that actively works to lower your monthly bills by preventing the energy from getting wasted in the form of hot steam.

To conclude, a normal pot is like a teacherless, unattended school classroom—noisy, chaotic, where everyone is shouting (vapor escaping). While a pressure cooker is a classroom where the teacher's in charge and everyone's behaving—every molecule is forced to stay in the room.

That's why it saves about 70% of fuel—it's not just faster, it's a closed loop system where no effort is wasted.

The pressure cooker also shows that boundaries aren't always a bad thing. They're like a password—necessary and shouldn't be shared with anyone. By sealing the lid and keeping our energy from leaking away into the kitchen air of distractions, we preserve the best part of ourselves.

Egypt

Egypt. When we think of this country, we think of a few 3-D triangles, barren desert, a rich teen whose tomb was never looted, mummies, and royalty. But the reality is nowhere near the dreamy royalty. Just one part of Egypt is all pyramids and static museums, the other half is like a factory that's been broken down and stalled.

One of the most advanced civilizations to once exist, who made iconic infrastructure, invented farming and irrigation, is now importing 60% of their food. They have the Nile River but have lost control, logic to run it properly. Today, they are dealing with high inflation and are stuck in a debt trap.

What went wrong? The ancient Egyptians, even the present ones, built the pyramids so well that they stopped asking, "What's next?" and instead went, "This is it."

Going back in history, Egypt missed iron. They were stuck in the bronze age, while everyone moved to iron. They had the best designs and hardware but had the wrong material to support it. They were trying to make 45-minute-long documentaries on wildlife in 2026, hoping to go against a

30-second "get-ready-with-me" reel —quite a daring move. Anyways, making iron requires a higher temperature than bronze. Egypt had the scorching sun but lacked wooden fuel and furnaces to compete with the iron age.

They moved from being a lab—creating new things, to being a museum—just holding onto old things. They spent all they had on the dead rather than the living.

This tells about how we need to shift from the old grids to Smart Grids. A grid is like a transport network for electricity. All the electricity made from power plants of coal, solar, wind, geothermal, and more is connected to the place where it's needed.

Simplifying it, Power plants → Phone charger. All the electricity is concentrated at one central power plant and

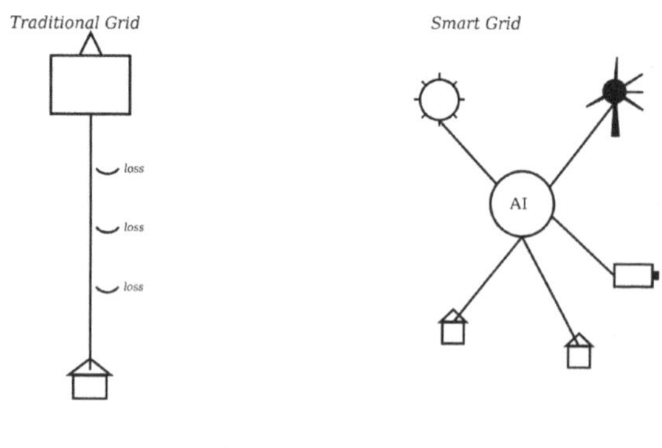

From museums to laboratories

flows down through big wires to your house. But it can't control the amount of electricity being pumped. It just pumps in massive amounts of energy, hoping it reaches the end.

It's just like a teacher shouting their lungs out in a free period, trying to maintain law and order—hoping someone bothers to listen.

But is it efficient? Well, it's like playing with dominos. If one domino falls in the middle, the rest are also doomed. There's no going back, leading to a total blackout. And this setup is the one currently being used widely.

Its transportation is like trying to climb a 30-feet-high mountain with a teaspoon of water—hoping we'll be left with the same amount by the time we reach the top. Well, obviously we end up losing a big part of it. That's exactly what happens with normal grids.

In a traditional grid, electricity must travel hundreds of miles from a power plant. As it travels through the copper wires, friction is created due to rubbing of electrons. This turns the electricity into useless heat. The wires at this point are just like bench warmers for birds to take a break on.

Whereas a smart grid uses distributed generation. The traditional grid just functions, the smart grid functions using a brain. It's a brain with sensors attached to every wire that measure every milliamp. It figures out exactly how much power is being made and where it's being sent at any point in time.

The smart grid doesn't set out on a long-long adventure to deliver electricity. It uses energy from the sources that are closest to where the power is needed. By reducing the distance and over supply of electricity, it cuts down the friction in the wires. The energy reaches the final point as useful power and not useless heat.

The smart grid isn't just the one with brains, but also a manager who manages the hybrid system and the MPPT—Maximum Power Point Tracking—to make sure no energy is wasted. To explain hybrid, it's a setup that doesn't just rely on one source. It combines solar + batteries.

At night or on a gloomy day, solar can't work. This is where the batteries come to the rescue. This is Energy Diversification.

While the MPPTs, as the name suggests, makes sure to squeeze out the maximum power out of a given source at any point. It's constantly adjusting to get the most output.

It doesn't take only what the sun gives—it's like using every bit of a face mask, serum, or a toothpaste nearing its end.

As it's said—"What you can measure, you can manipulate". It's given by Lord Kelvin who meant to say—if you don't track your cash flow and where you're spending your time, you can't improve your efficiency.

Egypt's horse view approach took a toll on them—instead of broadening out, they focused on one narrow thing, following the narrow path.

This is a kind of wake-up call for everyone following the traditional, single track railway path. Our learning system wants us to focus on one big exam for 365 days, spending our days and nights solving sums until the brain hits a blackout. That's what horse view is. If you trip once, your whole year—and your confidence goes down the drain.

We need to stop functioning like a domino with a single point of failure—dominos are best suitable as a game to be played on a random Sunday.

Being a smart grid, you can think of it like—If plan A fails, the alphabet still has 25 more letters. Along with the main, concrete plan A, you can have your B as finance, C as communication and D as history, or rather any permutations and combinations which will help one adapt and evolve with the rapid age of advancement.

Rather than going for a series circuit, why not be the parallel circuit. If the math circuit blows a fuse, don't blackout and go dark—the negotiation battery or the financial solar panels will kick in to keep the lights on.

Kellogg's Factory Visit

Corn Flakes—the famous breakfast cereal that's not just for breakfast. I've been on these crunchy dried flakes since I was a tiny human, and the best part? No clock-in time required. Whether it's dawn, dusk, midnight, or those odd hours during the night, these flakes are my go-to snacks. And you know what?

You spot Kellogg's literally everywhere—walk into any hotel, go anywhere in the world, and Kellogg's is the benchmark. At this point, Corn Flakes and Chocos are the breakfast MVPs that are always on the table. I mean, if they don't have it, it feels off—and would that even be an authentic breakfast spread?

But, apparently, those little dried flakes go through a rollercoaster loop-de-loop in an automated factory, all to get that red carpet makeover—going from dried corn to breakfast royalty. And I had a great opportunity to learn about this 'flake to flame' process at Kellogg's Taloja factory, witness real-time automation and the makeover of these cereals.

Walking into the factory, I saw some real Iron Man tech

with ginormous machines running everywhere. But, unlike the traditional, human labor operated factory bases, this one was fully automated—everything run and managed by machines themselves and humans only there for supervision and management.

At the factory, I was also briefed with the metal detection and robotic vacuum used in the segregation and packaging of the final products. My visit was an amazing learning experience thanks to Mr. Prashant—A great *imagineer* by profession with the instructional clarity of an experienced professor who made complex concepts super easy to grasp.

One thing I saw everyone put emphasis on was that 'input is everything'. Most of the time, people only care about the final product, but the essential part is to look at the starting line. If the input isn't perfect, the entire system is just a complicated way to fail.

Now, in a high-speed, automated factory environment like Kellogg's Taloja, the transition from raw grain to a perfect flake depends entirely on how the PLC manages the Input stage. In an automated factory, the input isn't just the corn, it's the data about that corn.

Before the machines even move, they receive signals from

sensing organs. The sensors measure the variable moods of the input such as moisture levels or temperature.

This was my first time hearing and learning about PLC. To explain, PLC—Programmable Logic Controller—is essentially the digital brain of the factory. While our brains handle thoughts and emotions, a PLC handles if/then rules at lightning speed to keep machines running with zero wasted movement.

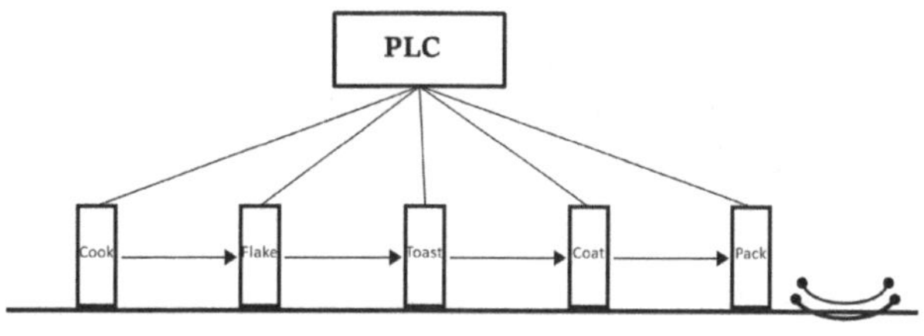

The supervisor that never blinks

For example, if the temperature is more than 205 degrees Celsius, then it lowers the gas flame. The PLC instantly sends an electrical pulse to the gas burner's motor. It turns the valve down, the flame shrinks, and the temperature drops—this is the output.

So, to summarize it, the PLC is the supervisor that never blinks and makes decisions within milliseconds.

Now, moving onto my favorite part—the smart use of the

domino effect in the factory. Usually when we talk about dominos, we think of a total crash in a designated order. But here it's the opposite. In the outside world, when one thing fails, everything crashes.

However, in the factory, if one tiny sensor detects an issue or a glitch, it doesn't cause chaos in the whole system. It's smart enough to stop right away, isolating the issue, so that the rest of the ingredients and mechanism stay safe.

In these massive, automated places like Kellogg's, the production line is a long row of standing dominos. Each stage depends on the one before it.

Cooking → Flaking → Toasting → Coating → Packing

Here, the PLC acts as the behind-the-scenes wizard—hovering between every single domino in that row, keeping them in check.

If the PLC—acting like eyes—senses that the first domino is leaning the wrong way or isn't aligned in the designated order, it reaches down and pauses the sequence before the second domino moves. By stopping the bad input at the start, it saves the energy of the entire factory.

In the end, without a PLC, a factory is just a 'complicated

way to fail', as the tiniest error at the start could end up in a chaos-loop. The PLC ensures that the dominos fall in the right way to ensure that the end result is a perfect bowl of corn flakes.

Other things which I was briefed with were the metal detection and robotic vacuum used in the segregation and packaging of the final products.

Now—for the part I love—their way of handling energy efficiency. Here, they don't waste even an ounce of energy. The engineers find solutions for problems that don't even seem like a problem unless one digs deep into it.

I learnt about their Steam Recovery where the hot exhaust is taken from the toasting ovens and, using heat exchangers, incoming air is prewarmed. It's a closed loop system where waste is just another word for unused potential.

To conclude, we usually need a driver to drive a car, without whom the car won't move forward. And even if you've done the same commute a million times, you still can't drive with your eyes closed—you must keep your eyes on the road and drive yourself.

But now, we have cars that don't have drivers—there are

driverless cars in China! And this is exactly what I observed in Kellogg's Taloja factory—a driverless factory. You just have to give the right input, and then it's automated—runs the whole mechanism and manufactures on its own.

I'll certainly look forward to seeing how many more amazing advancements Kellogg's will come up with.

R&D Lab Visit

Walking into the Research and development lab of AAK, I expected to see some dramatically explosive Iron Man tech and big, loud, flashy machinery.

In my case, I never knew what research and development labs were. Only that whatever was involved in production would have some intense, massive machinery. And to top it off, it was a lab that developed and manufactured customized ingredients for the bakery industry.

I thought baking was just adding flour, eggs, milk, and hope. But I guess my 'hope' ingredient wasn't enough. I've mastered adding hope but the baking part... not so much.

What struck me was the blend of innovation and practicality. In the labs, people were creating new innovations for their clients while being kind to the planet—a win-win situation according to me.

I, at times, feel like the world is obsessed with 'bigger'. In the world of energy, we spend billions trying to generate more power to overcome the distance of our inefficiency—

like making massive reactors, towering turbines, or continent spanning grids. These are the compensations we make for bad designs, cluttered and noisy minds.

Going through and learning about the R&D lab, I learnt how they were bridging the gap. Sometimes, the best thing you can do to have more power is not with some gigantic, Iron Man tech, but rather avoiding the power you already have from going to waste by simply standing in the right place.

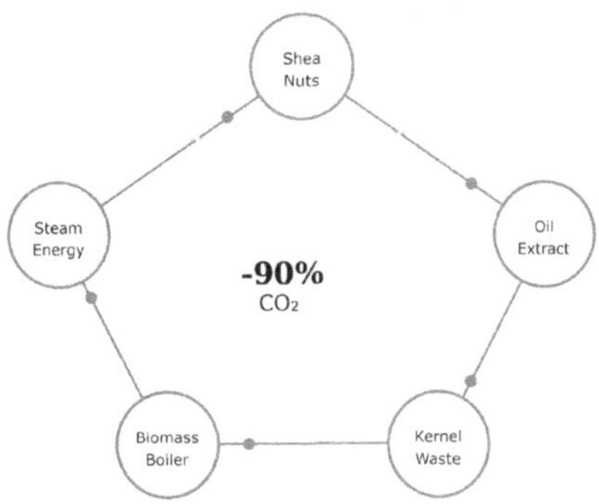

One nut powers the next

In some places, energy is treated like a drive-thru, fuel arrives from afar, gets burned—meaning wolfed down quickly—and the food wrapper is someone else's problem—meaning the waste is shipped off. But halfway through my visit, I learned that, at AAK's Aarhus site, they made a smart R&D proximity move by installing biomass boilers that run on shea kernel waste.

Meaning, they stopped looking for new fuel and started using the energy which was already sitting in a forgotten corner of their own factory.

So, what's shea kernel waste? You know the famously advertised shea butter lotion? Or just take shea butter or chocolate... to make these, nuts from the shea trees are crushed and two main things are left behind—shea shells, which is the hard outer layer, and shea meal, which is the dry, brown 'cake' left after squeezing out the oil.

For a while now, factories and local processors would just dump this out. But while others saw this as waste, AAK saw this as a battery. This waste is better than wood. Once dried, it actually burns better than traditional firewood, which also cuts down on cutting trees.

They feed the shea meal into giant biomass boilers, then these boilers burn the meal to create steam, which then spins turbines or provides the heat needed to process the next batch of nuts.

All this is basically a circular system where the nut itself provides the energy to turn itself into butter. At AAK's plant in Denmark, this move is slashing CO_2 emissions by 90%.

Now, moving back to the lab I visited, the lab doesn't just put products under the microscope and write a 20-page report on it, it builds them. They focus on specialty vegetable oils and fat having several applications with other varied items.

Who knew so much care and thought would be given to a little bar of chocolate or a piece of muffin. I loved seeing the confectionary and bakery lab where they engineer the mouthfeel of chocolates and even the puffiness of pastries.

While minimizing chemical input and maximizing crop protection, they don't just sell oil; they sell technology to make that oil more efficient, just like in energy efficiency!

After this visit, it turned out that you don't need an industrial park to innovate —just a small lab, all that tech packed into a space that looks like a regular office with some serious brains.

Budding Scientists

Budding Scientists, my first ever science project competition. Just so you're not mistaken, I never aspired to become a scientist. Growing up, I felt it was a tedious task just to learn everything someone broke their head over to discover.

So, I had no intentions of making something of my own. But now that I've found something that I like learning about, I love to dive deep into it, brainstorm, and come up with new solutions for new problems.

Anyways, getting back to the competition, it all started with me seeing the competition form and thinking of participating. Thanks to this, I was introduced to the topic 'energy efficiency' and decided to make my project on it.

The next task was to find group members who had good speaking skills since we had to present our project. A very easy task—I just gathered up my friends and asked if they were willing to do it with me and they agreed.

Before the actual competition, there would be selections for the ideas/topics that each house would present. Being my first time, I wanted to be prepared for anything the teachers asked me. So, I spent an entire day making articles, PPTs, and charts to convince the teacher to select my idea, just for no one else from my house to show up at all.

So basically, I was the only option my teacher had :) It was a little disappointing but I still proudly told everyone MY idea was selected.

However, the rest of the process wasn't as smooth. Since I was young, I've loved making models and projects, so it was not difficult to come up with them. The hard part was the execution.

Being someone who loves to rely on instincts, it was hard to coordinate with everyone—measure each and every cardboard being cut and attached. I spent almost

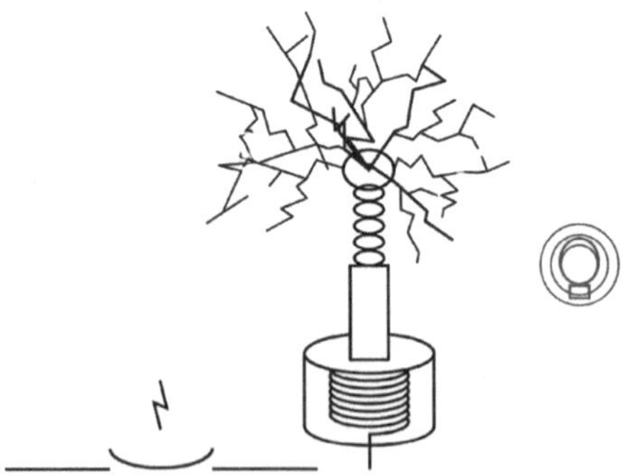

Where chaos meets discovery

two weeks surrounded by cardboards in my room while I worked on my science project- it was a cardboard filled chaos.

Just a few days before the competition, we came up with a new research topic on electrokinetic power generators. Imagine driving over a speed bump, and along with just slowing down, you're also helping in generating electricity!

Electrokinetic power generators are clever devices embedded beneath the road surface, capturing the kinetic energy of moving vehicles and converting it into electricity through electromagnetic induction.

All that was left to do was make the script and start practicing for the presentation. But who knew making a 3-minute-long script, with group members who had been so cooperative, would be so difficult.

At first, I made the entire script which everyone approved. But then, just two days before the exhibition, everyone had opinions of their own which led to us changing the script a dozen times—definitely quite frustrating. Though we somehow got past that, it wasn't the last speedbump we would experience.

A day before the exhibition, we were informed that we

would be scored based on both the teams from the same house (we had 6 houses and there were 2 teams from each house).

Now, since we didn't want our efforts to go wasted, I prepared a script for the other team and helped them with their models and their presentation too. But it turned out to be a fiasco since we were wrongly notified and, apparently, all teams were scored individually :(To top it off, we came to know this after the results were out.

It was now D-DAY—the day of the competition and exhibition. And it was truly a bumpy ride. I've got an incredible memory... in a bad way. It's super quick at fading away when I need it.

Even after going over the plan in my mind a dozen times, I forgot to get extension cords for our working model, and I realised it only after reaching the venue. Thankfully, we rang home and asked someone to get it to school immediately, having it before the competition started.

Then, as we thought that we were all set, just 10 minutes before the exhibition, the bulb from our other team's project broke. Their model was that of a Tesla coil, with the bulb—the heart of the project—lying shattered on the ground.

We thought that, as with the extension cords, we could get another bulb and rang home again. Just a minute or two before the event, we had the modern LED bulb. All set now, right?

But no. We actually needed a CFL bulb as it's the only one that could demonstrate the working of a Tesla coil. And before we could comprehend the problem and improvise, the judges came and our timer started.

Now, we decided to just go with the flow. They were the third team to go and we were the second—with everyone having 4–5 minutes to present and one minute for judges to ask questions.

A Tesla Tower, also known as a Tesla coil, works by using electromagnetic resonance to generate and transmit high-voltage electricity wirelessly.

Invented by Nikola Tesla in the late 19th century, the system consists of two main parts: a primary coil and a secondary coil, each tuned to the same resonant frequency. When electrical energy is supplied to the primary coil, it creates a magnetic field that induces a current in the secondary coil. This buildup of electrical energy results in extremely high voltages and can produce dramatic electrical discharges, like sparks or arcs.

After seeing the first team's presentation, I was pretty confident in ours since we were much more prepared. The minute we finished introductions and started our presentation, principal sir came up to us and we were basically giving him the presentation while the teachers went off to a corner.

Anyways, the presentation ended well and now it was time for questions. After hearing his question, I was at my wit's end. Apparently, we had a chart paper depicting various renewable sources and how it will take a long time for them to come into action.

Using that, we wanted to depict how energy efficiency can be implemented quickly and is a better alternative. But out of all the showy and great models we had related to the topic that I was well versed with, he decided to ask me a question from a part I hadn't laid eyes on. I felt like throwing that chart away at that moment.

But, as I got my mind together, I simply replied that I wasn't sure of the question and would surely look into it. This is exactly what my teacher taught us to say if we came across a question we weren't familiar with, so this save was all thanks to her.

Then it was time for our next team with the Tesla coil. And

honestly, even though that group was younger than us, I learnt a lot from them that day. Even after a major part of their model was broken, all of them kept their cool and handled the situation well. Thanks to their quick thinking, they could still demonstrate their model using the wrong bulb! How?

My team wanted to demonstrate the concept of energy efficiency through a model of the evolution of the bulb, so we had three bulbs attached to a wattmeter. The incandescent bulb, the CFL bulb, and the modern LED bulb. We thought of removing the CFL bulb after our presentation and giving it to them, but we couldn't get it out before their presentation started.

In the end, rather than getting the bulb out, the guy from the other team simply brought the coil near the CFL bulb and it worked! What an easy solution, right? Coming to think of it, it wasn't rocket science, just that everyone was out of their wits with nervousness, so it probably didn't click with anyone sooner.

Now, for the most important question—after all the chaos, did we win? Well... *no*. But do I regret any part of the process? Also no. It was a little disappointing when I heard the result but soon, it felt alright.

Even if I were to get a chance to go back and change anything, I wouldn't do it. Maybe it didn't click for the judges, but I ended up learning a lot during the process.

And that concluded our exhibition shenanigans.